PROJECT  SCIENCE

W9-BKN-836

Electricity and Magnetism

Terry Jennings

Illustrated by **David Anstey**

CONTENTS

SMITHMARK

MAGNETS

Long ago the people of Greece found a strange stone. It would pick up little pieces of iron, which stuck to it. It was called *lodestone*.

In the 16th century, William Gilbert found that if a piece of iron was rubbed with lodestone, the iron would pick up smaller pieces of iron. The piece of iron had become what we now call a *magnet*.

Today magnets come in all shapes and sizes. Nearly all of them are made from iron or steel. Steel is mainly made from iron.

FIND THE MAGNET

Many everyday household objects have magnets in them. Look at the pictures below. Where is the magnet in each object? What is it used for?

Are there any magnets around your home? If so, what are they used for? Can you think of any other places where magnets are used?

Make a list of all the things in your home and school that use magnets.

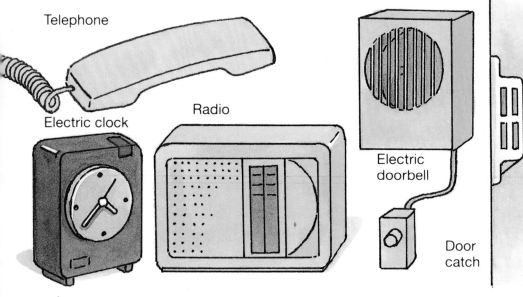

Telephone

Electric clock

Radio

Electric doorbell

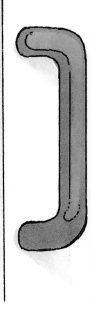

Door catch

WHAT WILL MAGNETS ATTRACT?

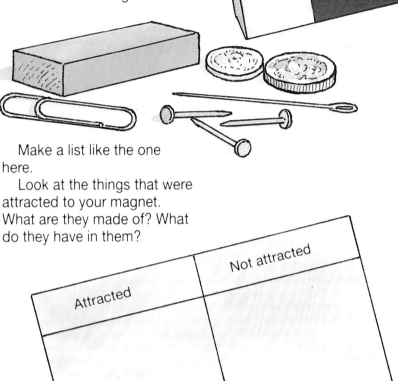

When a magnet picks up a pin or a small nail, we say these things are *attracted* to the magnet. The magnet affects the area around it, pulling *magnetized* objects toward it. This area is known as the *magnetic field*.

Find out what other things a magnet can attract by doing the next project.

1 Put your magnet near each object in turn. Which are attracted to the magnet?

Make a list like the one here.

Look at the things that were attracted to your magnet. What are they made of? What do they have in them?

You will need: a magnet; coins; tacks; plastic comb; pieces of wood; glass; paper clip; sewing needle.

Attracted	Not attracted

Magnetic Earth

If the Earth were cut in half, we could see that the core (center) is made of metals that are mainly iron. These metals act like a huge magnet, giving the Earth a magnetic field around it.

Like a magnet, the Earth has a south pole and a north pole (see page 10).

You can see a magnetic field for yourself.

1 Place a piece of paper over a magnet.

2 Sprinkle iron filings onto it. Tap the paper.

The filings form a pattern showing the magnetic field. Why are there more filings around the ends of the magnet? You can find out on page 9.

MAGNETIC FORCE

Most magnets are made of the metals iron or steel. The objects attracted to magnets are also made of iron or steel, or they have iron and steel in them.

Some magnets attract objects more easily than others. We say that they have a stronger *magnetic force*. They also keep their magnetic strength longer. This is because they are made of special kinds of steel or other magnetic metals. Cobalt and nickel are also magnetic.

Some magnets are strong, others are weak. Find out how strong your magnet is by doing the experiment shown below.

MAGNETIC CHAINS

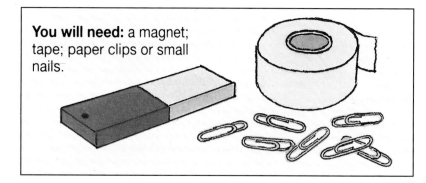

You will need: a magnet; tape; paper clips or small nails.

3 Now hang your paper clips 3/4 of an inch (2 cm) from the end of the magnet. How many can you get to hang in a line?

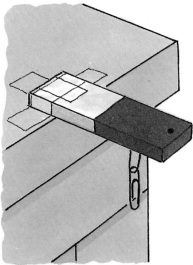

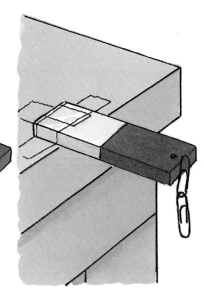

1 Put the magnet at the edge of the table. Fix it firmly at one end with the tape.

2 Hang paper clips on the tip of the magnet. How many can you get to hang in a line?

Keep trying, each time moving the paper clips farther away from the end of the magnet. What do you notice?

Have a contest with your friends. Use the same paper clips or nails each time. Who has the strongest magnet?

MAKE A PIN JUMP

Another way to test the strength of a magnet is to see how high it can make a pin jump.

1 Put a newspaper on a table and stand a brick on the top. Tape your magnet to the top of the brick.

2 Lay a pin on the newspaper underneath the magnet. Does it jump up? If not put the pin on a book and keep adding books until it does.

3 Measure how high the pin can jump.

MAGIC MAGNETS

We have seen that magnetism can go through the air. What else does it pass through? You can test this for yourself.

You will need: a magnet; thick cardboard; plastic lid; thin piece of wood; glass; paper clips; small nails.

1 Put your paper clips and nails on the cardboard.

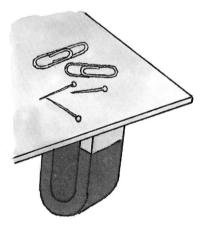

2 Move the magnet around underneath. Do the metal objects move?
   Try again using the lid, the glass, and the piece of wood. Does magnetism pass through them all?

"Tricky Clips"

Try a trick on your friends.

1 Put a paper clip in a glass of water. Ask your friends if they can remove the paper clip without touching the glass or getting their fingers wet.

2 Now it is your turn. Hide the magnet in a piece of "magic" cloth. Hold it very close to the glass near the paper clip.

3 Slowly move the cloth from the bottom of the glass to the top. Watch the paper clip rise as if by magic.

MAKING A MAGNET

When objects are attracted to a magnet they become magnetized. This means that they can also act like magnets and attract iron or steel objects.

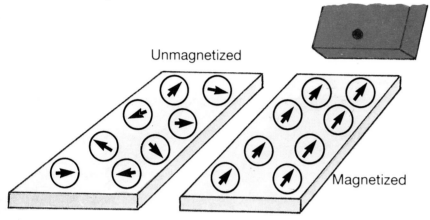

Unmagnetized

Magnetized

When a piece of iron or steel touches a magnet, it becomes a magnet itself. A piece of iron or steel is made up of millions of tiny magnets pointing in all directions. When a large magnet is near them, all the tiny magnets point in one direction – the iron or steel is magnetized.

NEW MAGNETS FROM OLD

If you already have a magnet, it is quite easy to make another one.

1 Lay the nail on the table and carefully stroke it with your magnet 20 times. Each time stroke the nail with the same end of the magnet and in the same direction.

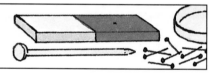

You will need: a magnet; large nail; plastic lid; pins.

2 Fill the lid with pins. Put your nail near the pins. Will it pick any of them up? If so, how many? Is your nail a magnet?

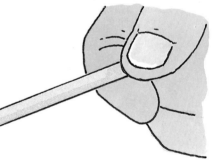

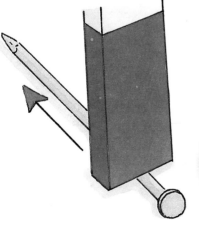

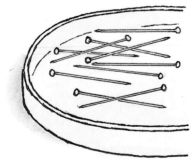

Keep stroking your nail with the magnet. Can it pick up more pins than before?

You have made a magnet from a steel nail. What else can you make into a magnet? It would be useful to have a screwdriver which is a magnet. Can you make one?

DANCING BUTTERFLY

This is a fun way of testing your new magnet.

You will need: paper; colored pens or paints; poster board; tape; paper clip; scissors.

1 Draw a large picture of a butterfly on a sheet of paper. Color it with paints or crayons.

2 Cut out the butterfly and tape a paper clip to its underside.

3 Rest the butterfly on the poster board. Hold a magnet out of sight under the board right beneath the butterfly. Move the magnet and the butterfly will move.

Ask you friends to guess how your magic butterfly works.

Magnetic Attraction

Is the magnetic attraction of a magnet equally strong all over? You can test this very simply.

1 Sprinkle pins or tacks evenly over a sheet of newspaper and roll your real magnet in them. Where do most pins or tacks stick? Is it the ends of the magnet, or is it the middle? Or are the pins or tacks stuck all the way along the magnet?

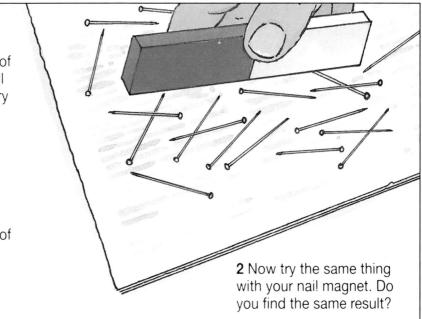

2 Now try the same thing with your nail magnet. Do you find the same result?

MAGNETIC POLES

Have you noticed that nails, tacks, and paper clips stick best to certain places on your magnet? These places are called the *poles* of the magnet.

Where are the poles of the magnets you have been using? There are two poles on each magnet. One end is called the north pole. Often this is marked or colored. The other end of the magnet is called the south pole.

South pole

North pole

TOGETHER OR APART?

Something unusual happens when two magnets are brought near to each other? Test this for yourself.

You will need: plastic lid; 2 magnets; bowl of water.

1 Float the lid on a bowl of water.

2 Rest a bar magnet on the lid and wait until the lid has stopped moving.

3 Slowly bring the north pole of another magnet near to the north pole of your floating magnet. What happens?

4 Bring the south pole of a magnet near to the south pole of the floating magnet. Does the same thing happen?

5 Finally, bring the north pole near to the south pole of the floating magnet. Then the south pole near to the north pole of the floating magnet. What happens in both tests?

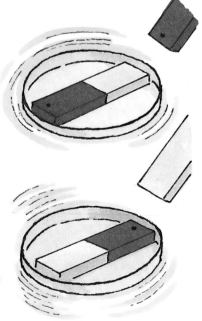

As long as something is made of iron or steel it is easy to make into a magnet. When you make a new magnet, you do not make the first one any weaker. All the energy to make the new magnet comes from your muscles.

Why do magnets become weak? Find out for yourself by doing the experiment described below.

LOST POWER

You will need: a large nail; magnet; block of wood; small nails.

1 Magnetize the nail by stroking it with a magnet. Test how strong your new magnet is by counting how many small nails or tacks it can pick up.

2 Bang your new magnet very hard against a block of wood. Do this ten times. Or drop it on a hard floor ten times. Then test its strength again. Has there been a change?

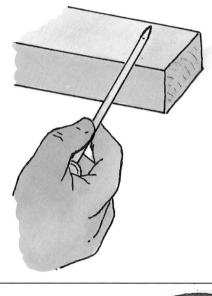

What happens if you bang or drop the magnet ten more times?

This shows you should always handle your magnets carefully. Heating magnets will also weaken them. Never put them near a naked flame.

Taking Care of Magnets

Magnets get weaker just lying around. When you buy a magnet it usually has a piece of iron called the keeper with it. The keeper helps the magnets to stay strong. If you lose the keeper, use a nail instead. Store your magnets like this to keep them strong.

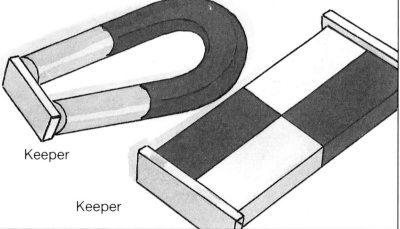

Keeper

Keeper

MAGNETIC TOYS

If you bring two north or two south poles together they will *repel* or push each other away. A north and a south pole or a south and a north pole will attract or pull toward each other.

Both north and south poles will attract objects made of iron and steel. We can use the fact that magnets can push and pull to make magnetic toys.

DRIVING TOY CARS

You will need: a large sheet of poster board; colored pens or pencils; 2 small plastic cars; 2 nails; 2 short sticks; 2 magnets; tape; books.

2 Tape a nail underneath each car and put them on your model roads.

3 Tape a magnet to each stick. Hold the sticks under the poster board.

4 Use them to steer the cars along the roads. You can have races with a friend.

1 Draw some roads on your poster board. Make lots of crossings and junctions. Rest the board on four piles of books. Make sure that you can get your hand underneath.

FISHING GAME

You will need: poster board; colored pens or paints; scissors; paper clips; 2 popsicle sticks; thread; 2 magnets.

1 Cut out a few poster board fish shapes and color them. Fix a paper clip to each one. Write a different number on each fish.

BOATING GAME

2 Tie a thread to the end of each stick. Tie a magnet on the other end of each thread. You now have two magnetic fishing rods.

3 You and a friend can take turns trying to catch a fish. Add up the numbers on all the fish you catch. The person who has the highest score is the winner.

You will need: several corks; thumb tacks; paper; small pins; tape; bowl of water.

1 Ask an adult to cut the corks in half. Stick a thumb tack in the curved side of each cork.

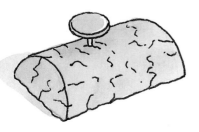

2 Stick a pin in the middle of the flat side of each cork. Cut out small paper triangles and tape them on to the pins to make sails.

3 Float the boats in a bowl of water. Use your magnetic fishing rods to steer the boats around the bowl.

You can have boat races with your friends.

FINDING THE WAY

Long ago, the ancient Chinese discovered that a piece of lodestone, hung from a string, always pointed north and south. They had made a simple *compass*.

Compasses are used to help people find their way. Ships and airplanes have compasses to help them find their way across the sea and sky. People walking in the mountains and across deserts also need to use compasses. Find out below how to make a simple compass for yourself.

THIMBLE COMPASS

You will need: a magnet; large cork; modeling clay; plastic thimble; 2 needles.

1 Put a needle into the cork and balance the thimble on the end of it.

2 Magnetize the second needle. Using the modeling clay, fix it to the top of the thimble.

3 Stand in your yard and find out what buildings are to the north, south, east, and west of your compass.

Special Boxes

Bring a piece of steel near your homemade compass. What happens? Does it still point north? Why do you think real compasses have boxes made of brass, wood, or plastic?

FLOATING COMPASS

You will need: a magnet; large cork; saucer of water; large sewing needle; 4 adhesive labels.

1 Ask an adult to cut a thin slice of cork and float it on the saucer of water.

2 Magnetize the needle and gently rest it across the piece of floating cork.

One end of the needle will point north. The other end will point south. Ask an adult to tell you which is which, or check with a real compass.

3 Write out N (for north), S (for south), E (for east) and W (for west), on four little adhesive labels. Stick them around the saucer in the right places.

4 Use your compass to tell you which direction your school is from your home.

TREASURE HUNT

1 Draw a map of your yard and mark where you are going to hide the treasure and the clues leading to it. Mark your starting point (X).

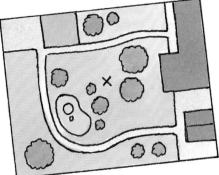

2 Stand at your starting point with a compass. Turn toward the place where you have decided to hide a clue. Write down the direction on a piece of paper – this will be your first clue.

3 Walk towards the hiding place counting the number of paces. Add this to the first clue. Mark the spot – this is where you will hide the second clue.

4 Using your compass find the direction of your second hiding place. Pace out the distance. Write down the direction and number of paces – this is your second clue. Hide it in the first hiding place.

5 Repeat this to find out the direction and position of the treasure – this will be your third clue. Hide it in your second hiding place. Don't forget to hide the treasure!

6 Give the first clue and the compass to your friends. Stand them at the starting point. How long does it take them to find the treasure?

MAGNETS AND ELECTRICITY

There is another kind of magnet which uses *electricity*. It is called an *electromagnet*. An electromagnet will only work while the electricity is turned on. It loses its magnetism when the electricity is turned off.

You can see below a few different objects, found in homes, offices, and factories, that use electromagnets.

MAKE YOUR OWN ELECTROMAGNET

You will need: a large bolt; thin plastic-covered wire (39 in; 100cm long); battery (4 1/2 volt); dish of tacks.

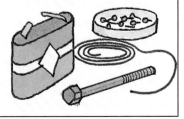

1 Wrap the wire, neatly and evenly, around the bolt about ten times.

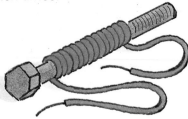

Ask an adult to strip the plastic covering off both ends of the wire.

2 Join the two ends of the wire to the battery.

3 Hold one end of the bolt in a dish of tacks, then lift it up again. How many tacks are clinging to it?

4 Undo the wire from the battery. Now wind ten more turns of the wire around the bolt. Rejoin the wire to the battery. Test the bolt again.

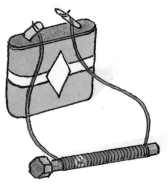

How many tacks will your electromagnet pick up now? Keep winding more wire and testing the strength of your electromagnet. What do you discover?

If you want to make a switch for your homemade electromagnet, you can find out how to on page 21.

MAKE A MODEL MERRY-GO-ROUND

You can use your
electromagnet in this model.

You will need: the lid from
a pint ice cream carton;
small tube cap; large cork;
large sewing needle or short
piece of stiff wire; paper
clips; pliers or gloves; glue;
colored pens.

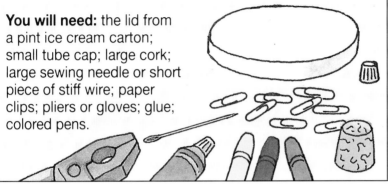

3 Hold your electromagnet
near one of the paper clips.
Turn on the electricity as the
clip goes by. Keep doing this
as each paper clip comes
near.

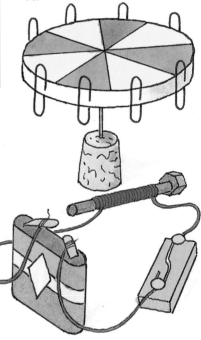

1 Using the gloves or pliers,
push the needle into the
middle of the cork.

2 Glue the tube cap into the
middle of the lid. Color the
other side of the lid, and put
the paper clips evenly all
around the edge. Balance the
lid on the needle.

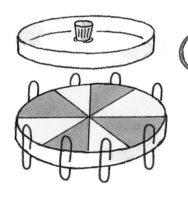

You will soon find you can
keep your merry-go-round
turning.

Cranes

Junkyards use
electromagnets for lifting
and sorting metals. The
electromagnet is on a
special crane. When the
electricity is turned on, the
crane picks up pieces of
iron and steel. When the
electricity is turned off, the
crane drops the metal.

ELECTRICITY

Most of the electricity we use in our homes, schools, shops, and factories is made in a large building called a *power plant*.

You can see below how the electricity reaches the buildings in your area. In the power plant fuel is used to heat water and turn it into steam. The steam pushes around a big fan, called a *turbine*. The turbine turns a machine called a *generator*, which makes electricity. The generator is something like a large *dynamo* which works the lights on some bicycles. You can see how it does this on page 28.

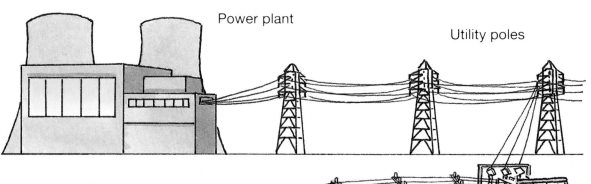

Power plant

Utility poles

From the power plant the electricity is carried along thick wires, or *cables*, until it reaches our homes or other buildings. The cables are carried overhead by *utility poles* or buried under the ground.

Underground cable

We use this electricity to work many things. Nine of them are shown in the pictures below. Can you say what they are?

Batteries also make electricity. You can find out how they do this on the next page.

NEVER play with plugs, sockets, wires, or electricity as they could kill you.

Batteries

Make a collection of batteries. Used ones will do. Look at them carefully. Each battery has two *terminals*. The terminals are the places where electricity flows from the battery and back into it.

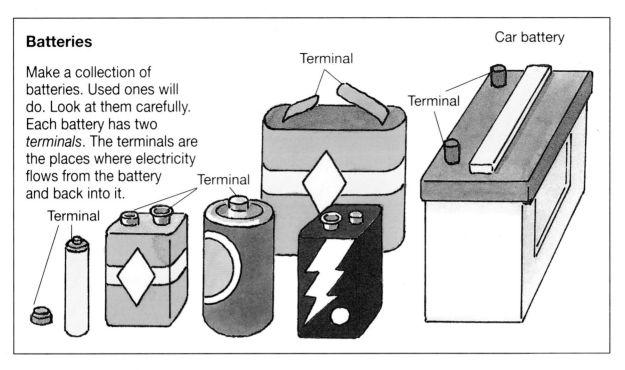

Terminal

Car battery

Terminal

Terminal

Terminal

Terminal

FLASHLIGHTS

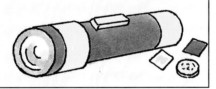

You will need: a flashlight; poster board; piece of thin plastic; coin.

Flashlights get their electricity from batteries. Guess how many batteries your flashlight has. Look carefully at a flashlight. Switch it on. Does the bulb light?

1 Take the flashlight apart. Can you see all the parts shown in the picture? Did you guess the correct number of batteries?

2 Put a small circle of poster board between the batteries. Switch the flashlight on. Does it light?

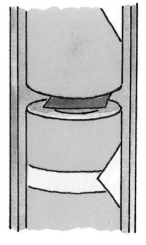

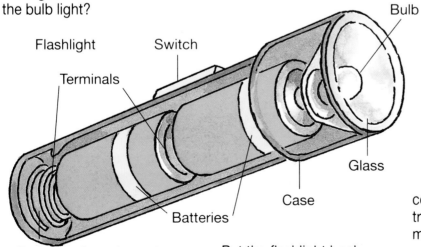

Flashlight

Switch

Terminals

Bulb

Glass

Case

Batteries

Spring (to keep batteries in place)

Put the flashlight back together again. Make sure that it still works.

Try putting a small copper coin between the batteries. Or try a piece of plastic or other materials.

You could make a list. Which of them let the bulb light? Which of them do not?

19

ELECTRICAL CIRCUITS

Electricity flows along wires something like water flowing through a pipe. To make electricity come out of a battery it needs a path, such as a wire, to travel along.

The wire must go from one terminal of the battery to the other. This loop of wire is called a *circuit*. If there is a gap or break in the circuit the electricity cannot flow.

LIGHT THE BULB

Can you make a flashlight bulb light with just one piece of wire and a battery? The picture shows you how.

Terminal

Terminal

You will need: a bulb in a bulb holder; battery; 2 pieces of plastic-covered wire.

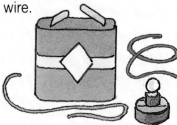

Make sure that the two ends of both wires are bare.

1 Wind a wire around each of the battery terminals.

2 Fix the other ends of the wires to the screws in the bulb holder. Does the bulb light?

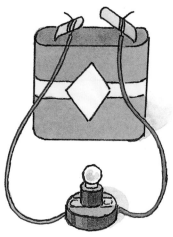

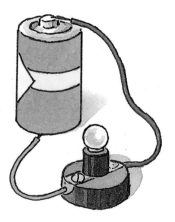

You can find out in the next project how to keep the two ends of the wire touching the bulb and the battery, so that you do not need to hold them.

ON/OFF

To turn the electricity on and off easily, and save the power in your battery, you can make a *switch*.

You will need: a small block of wood; 2 thumb tacks; 2 pieces of plastic-covered wire; paper clip.

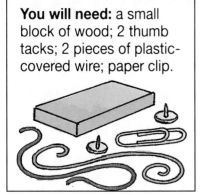

1 Press the thumb tacks into the wood a little way.

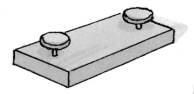

2 Wind the bare end of each wire around the thumb tacks.

3 Bend the paper clip around one of the thumb tacks.

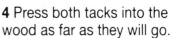

4 Press both tacks into the wood as far as they will go.

You can now attach the other ends of the wires to your battery in the usual way.

5 To complete the circuit, push the paper clip down so that it touches the other tack.

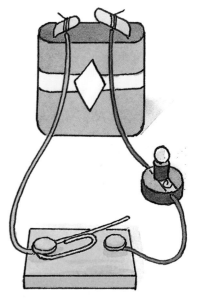

Make a Burglar Alarm

You can change the switch you have made to make a simple burglar alarm.

1 Bend a piece of thick tin foil upward.

2 Put the switch under a mat near a door. When someone steps on the mat, the tin foil will press down on to the thumb tack completing the circuit and the bulb will light. Make sure that when you place the bulb inside the room it is well away from the door.

CONDUCTORS AND INSULATORS

Some materials let electricity flow through them easily. They are called *conductors*. Other materials stop electricity from passing through them. They are called *insulators*.

Electrical screwdriver

Find out which materials are conductors and which are insulators by doing the project below.

Metal blade

Insulated handle

START AND STOP

You will need: a block of wood; 2 thumb tacks; bulb in a holder; 3 pieces of insulated wire; battery; screwdriver.

1 Put two thumb tacks in a block of wood and join the wires to them. Join the wires to a battery and bulb holder.

You now have a circuit but there is a gap between the two thumb tacks. And so the bulb will not light.

2 Lay the blade of a screwdriver across the two thumb tacks. What happens?

Now lay the handle of the screwdriver across the drawing pins. Does the bulb light?

ALWAYS make sure your connections are tight.

3 Lay things made of wood, metal, glass, plastic, cloth, and any other materials you can find, across the drawing pins. Which are insulators?

Make a list of insulators and conductors.

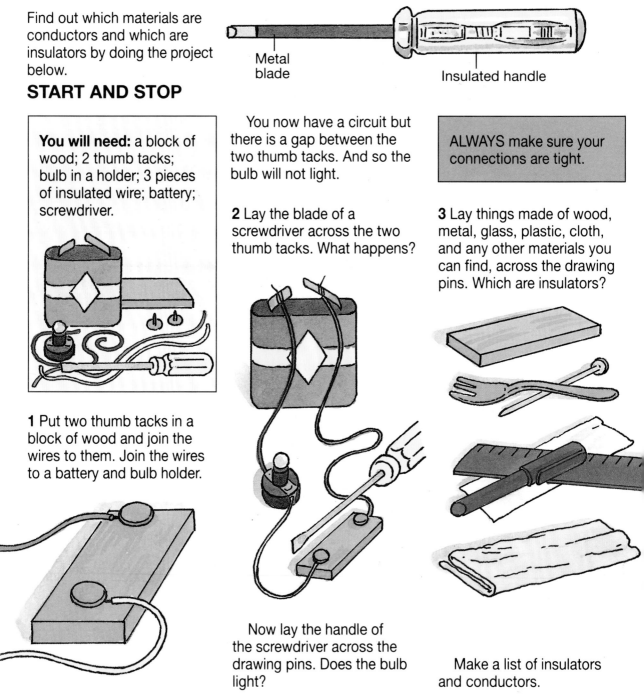

A DIMMING SWITCH

Some materials do not completely stop electricity from passing through them. One of these is pencil lead, which is made of graphite. You can use pencil lead to make a dimming switch.

You will need: a battery; bulb in a holder; old pencil; 3 pieces of insulated wire; paper clip.

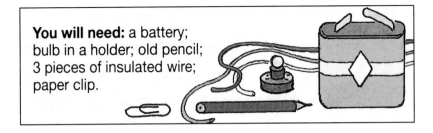

1 Join the wires to the battery and bulb holder.

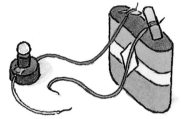

2 Ask an adult to cut half the wood away from a pencil.

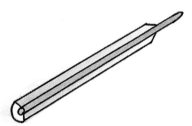

3 Fix one end of the wire to the pencil lead with a paper clip. Press the other end of the wire on to the pencil lead near it. Does the bulb light? Is it bright or dim?

4 Now move the end of the wire farther away and press it down on the pencil lead. What happens?

Live Wires

Conductors are used to carry electricity to places where it is needed. The insides of wires and cables are conductors. We use insulators to stop electricity from leaking into places where it is not wanted. Wires and cables have covers which are insulators. So do plugs and sockets.

The insulators on utility poles are made from a material like pottery to stop the electricity from leaking out and injuring people.

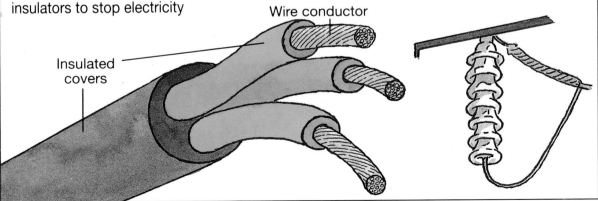

Wire conductor

Insulated covers

BRIGHT LIGHTS

Sometimes we use more than one bulb at a time. Christmas tree lights consist of a whole line of small bulbs sharing the same electrical plug which goes into a socket in the wall.

Find out what happens when we use more than one bulb or battery in a circuit.

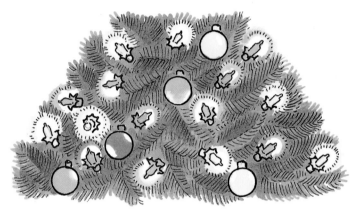

STRINGS OF LIGHT

You will need: 3 bulbs in holders; 4 pieces of insulated wire; battery.

1 Make a circuit with just one bulb in it. Does the bulb light? Is it bright or is it dim?

2 Add another bulb to the circuit, using a third piece of wire. Do the bulbs light?

3 Try adding a third bulb. Are they brighter or dimmer than when you had only one bulb?

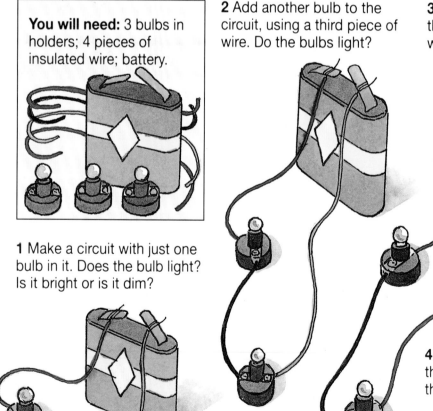

4 Carefully unscrew one of the bulbs. What happens to the others?

Do you know now why, when one of the lights on a Christmas tree goes out, the others may go out as well?

TRAFFIC LIGHTS

When bulbs all share the same power from a battery they give out only a dim light. If you take out one bulb, you break the circuit and the other bulbs go out as well.

There is another way to join up your bulbs to the battery so that each bulb has its own circuit. It is called a parallel circuit. We can use this circuit to make traffic lights.

You will need: 2 pieces of wood; battery; 7 pieces of insulated wire; 3 bulbs in bulb holders; 6 screws; wood glue; 4 thumb tacks; 2 paper clips; paints.

---

1 Ask an adult to screw the bulb holders on to a piece of wood. Glue the second piece of wood to the first.

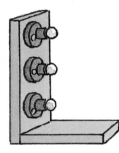

2 Paint the light bulbs red, yellow, and green.

3 Make a double switch using 4 thumb tacks. Straighten one end of each paper clip. Wire the switch and the lights to the battery as shown.

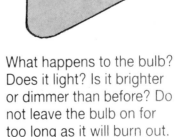

4 Turn each switch on. Do all the lights work? Can you use them like traffic lights?

Extra Power

What will happen if you make a circuit with two batteries in it? Try it and see.

1 Make a circuit with just one battery in it. Notice how bright or dim the bulb is.

2 Now add another battery. Make sure your two batteries are facing the right way.

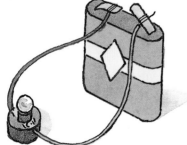

What happens to the bulb? Does it light? Is it brighter or dimmer than before? Do not leave the bulb on for too long as it will burn out.

25

HOMEMADE LIGHT

Light bulbs work by the electricity flowing through them. The electricity heats the *filament* so that it glows white hot. The glass bulb is filled with a gas which helps to stop the filament from burning away too quickly.

Another type of light consists of a long tube. In this type of light, electricity passes through special gases instead of wire filaments. The gases give off rays which strike the chemical coating on the inside of the tube making it glow.

Filament

You can test how electricity heats a bulb or an electric heater for yourself.

ELECTRIC HEAT

You will need: a square of wood; 2 nails; hammer; thin copper wire (39 in; 100 cm); battery; 3 insulated wires; switch.

2 Wind the thin wire around a pencil neatly and evenly with no overlap.

3 Gently slide the pencil out so that the wire forms a coil like a spring.

4 Join this coil across the gap between the two nails.

5 Join up the rest of the circuit, with a battery, switch, and wires.

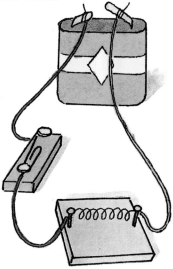

1 Hammer the two nails into the wood.

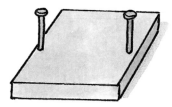

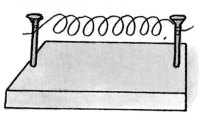

Switch the electricity on. Watch carefully. What do you notice? Put a finger near the coil. DO NOT TOUCH THE COIL. What do you notice?

ACID POWER

We can make electricity from chemicals like those found in a flashlight battery.

As the amount of electricity we can make is small, it will not light up a flashlight bulb, so we need to make an electric *current* detector.

You will need: a very small cardboard box; compass; thin insulated wire; battery; lemon; piece of zinc metal; copper coin.

1 Place a small magnetic compass in the box.

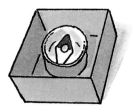

2 Wrap the wire neatly and evenly around the box 30 to 50 times.

Leave room at either end so that you can still see what the compass needle does.

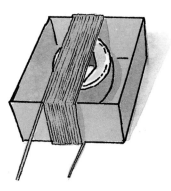

3 Join up your detector to a flashlight battery. Watch what happens to the compass needle. It will swing back and forth.

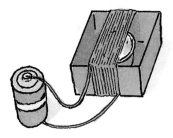

4 Make two grooves in a lemon skin. Push the copper coin into one slit and the piece of zinc metal into the other. Make sure the metals are not touching each other.

5 Hold the bare end of one of the wires from your current detector against the copper coin. Hold the bare end of the other wire against the piece of zinc.

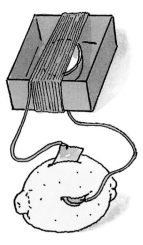

Watch what happens to the compass needle. Is an electric current produced?

Try the same thing using a potato instead of a lemon. Or, put the two metals in a little jar containing vinegar. What happens?

Fuses

These are thin wires fitted into the electrical machines, plugs, and fuse boxes in our homes. If something goes wrong, the thin fuse wire melts. It makes a gap in the circuit so that the electricity can no longer flow.

A circuit breaker is a safety switch. If a fault occurs it switches off the current instantly. It can form part of a plug or socket, or replace the fuse in a fuse box.

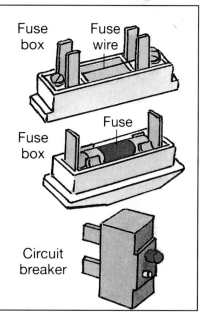

Fuse box   Fuse wire

Fuse

Fuse box

Circuit breaker

MAKING ELECTRICITY

One of the most important discoveries ever made was in 1831 by Michael Faraday. He found that when a magnet is moved near a coil of wire, an electric current is made. This led to the invention of the dynamos and generators which make our electricity.

You too can make electricity like Michael Faraday.

COILED ELECTRICITY

You will need: a cardboard tube (4in; 10 cm long); 2 cardboard disks; thin insulated wire; current detector; bar magnet; compass.

1 Glue a disk to each end of the cardboard tube to make a reel.

2 Wind 200 turns of the wire evenly on to the reel.

3 Join the coil to the current detector.

4 Move a strong bar magnet in and out of the coil. Watch what happens to the compass needle. Does the needle change direction when you push the magnet in and out of the coil?

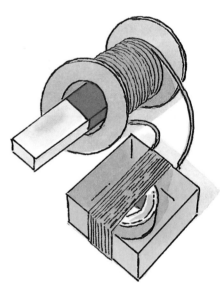

A Bicycle Dynamo

On many bicycles the lights are powered by a simple dynamo or generator. When the bicycle wheel goes around, it makes a magnet turn inside a coil of wire inside the dynamo. This makes electricity flow so that the lights work.

If you watch someone cycling with a dynamo you will notice that the faster they cycle the brighter the light is. What happens when the person stops cycling? Why could this be dangerous on a dark night or in foggy weather?

ELECTRIC MOTORS

An electric motor turns electrical energy into movement energy. You can make a simple electric motor for yourself.

You will need: thin covered copper wire (39in; 100 cm); 2 paper clips; rubber band or tape; battery; magnet; thick pen; modeling clay.

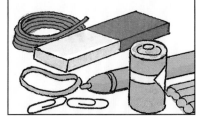

1 Carefully wind the wire around the pen. Leave about 2 in (5 cm) of wire free at each end.

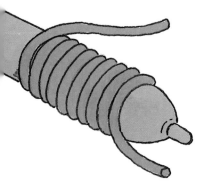

2 Remove the pen and loop each free end of the wire once or twice through the coil to stop it from unwinding.

3 Straighten out the two ends of the wire and ask an adult to strip the covering off them.

4 Bend the paper clips and fix them to the ends of the battery with a rubber band or tape.

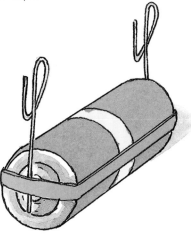

5 Stick a lump of modeling clay on either side of the battery to stop it from rolling.

6 Rest the coil over the paper clips. Make sure that it is level.

7 Hold one end of a magnet near the coil and spin the coil gently. It should continue to spin.

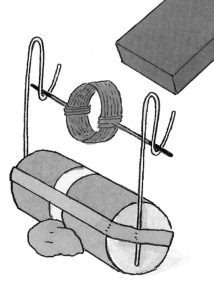

Electric motors are very clean and quiet. They can be easily turned on and off. Look around your home or school. Where can you see electric motors being used? What work do they do?

STATIC ELECTRICITY

There is another kind of electricity called *static electricity*. It is made when certain materials are rubbed together. One kind of static electricity which you will have seen is lightning. This huge spark of electricity is made when drops of water and ice in a thundercloud rub together.

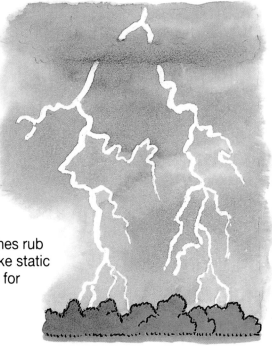

Have you ever seen tiny sparks or heard little crackles when you take your clothes off? This static electricity is made when your clothes rub together. You can make static electricity quite easily for yourself.

STICKY BALLOONS

Can you make a balloon stick to the wall without using any glue? All you have to do is to rub the balloon several times on a woolen sweater. Then hold it against the wall. Static electricity will make it cling to the wall.

1 Tie two balloons together with a length of thread. Rub both balloons with a cloth.

2 Loop the string over a hook, so that the two rubbed sides of the balloons are near to each other. What happens?

You can try it with two balloons.

30

"ELECTRIC" COMBS

You will need: a comb; tissue paper; woolen cloth.

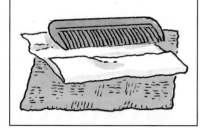

1 When your hair is clean and dry, comb it hard with a plastic comb.

Now watch in the mirror and bring your comb near a loose strand of hair. Does the hair rise up?

2 Tear up some tissue paper into tiny pieces. Rub a plastic comb hard with the woolen cloth (or the sleeve of a woolen sweater). Then hold the comb close to the pieces of paper. What happens to them?

3 Rub a plastic comb with a piece of woolen cloth. Hold the comb close to a thin stream of tap water. Watch carefully. What happens?

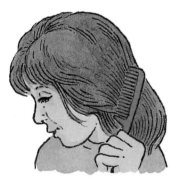

These experiments work because when the comb is rubbed it becomes charged with static electricity. This attracts the hair, the pieces of paper, and the stream of water.

Static Materials

1 Collect together materials such as paper, wood, rubber, glass, and various kinds of plastics. Make sure everything is really dry.

2 Rub each material with a woolen cloth in turn. Then see if it will attract little bits of tissue paper.

Which materials make static electricity the best?

TOYS AND GAMES

You can have fun making models and games which use electrical circuits.

MAKE A MODEL LIGHTHOUSE

You will need: a bulb in a bulb holder; poster board (6 in; 15 cm wide); 2 pieces of insulated wire (12 in; 30 cm long); battery; switch; modeling clay; tape.

Batteries will last much longer if you turn off the bulbs when you are not playing with them. If a battery is worn out, throw it away or it may leak and damage your models.

1 Roll the piece of poster board around the bulb holder and fix it with tape.

3 Stand the tube upright and press modeling clay around the base.

4 Join the wires to a battery and switch.

2 Attach the bare ends of the wires to each screw on the bulb holder. Tape them to the poster board tube.

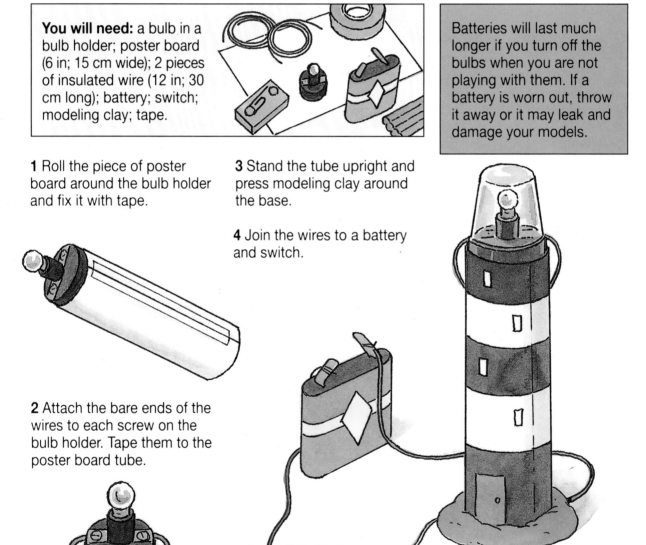

You can paint your lighthouse and cover the bulb holder with a small jar, or a clear plastic cup to make it look more like a real one.

THE STEADY HAND GAME

Make this game. Then test your friends to see who has the steadiest hand.

You will need: 2 pieces of florists' wire (20 in; 50 cm and 12 in; 30 cm long); 3 pieces of insulated wire (two of 12 in; 30 cm and one of 8 in; 20 cm long); bulb in a bulb holder; battery; piece of wood; nails or tacks.

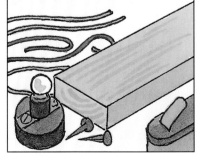

1 Bend the shorter piece of florists' wire to make a small loop at one end.

2 Bend the other piece of florists' wire into a wavy line. Try not to get any sharp kinks in it.

3 Thread the wavy piece of wire through the loop.

4 Nail or tack each end of the wavy line onto a piece of wood. Leave an inch or two free at each end.

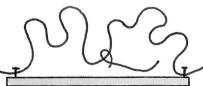

5 Join the bare end of a long piece of insulated wire to one end of the wavy piece of wire (**1**).

6 Join the other end of the covered wire to one of the screws on the bulb holder.

7 Join up the loop of wire to one terminal on the battery using a long piece of wire (**2**). Link the battery and bulb holder by the shorter piece of covered wire (**3**).

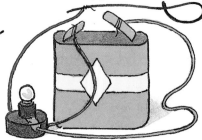

8 Touch the wire loop against any part of the wavy piece of wire. The bulb should light.

The game is now ready. See if any of your friends can move the loop of wire from one end of the wavy line to the other, without making the bulb light.

GLOSSARY

Here are the meanings of some words which you might have met for the first time in this book.

ATTRACTED: when two or more things are drawn together by a force we say that they are attracted.

BATTERY: a small device for making and storing electricity.

CABLE: a bundle of wires protected by a plastic tube.

CIRCUIT: the path taken by electricity around a series of wires and connections. If there is a break in the circuit the electricity will not flow.

COMPASS: an instrument with a swinging magnetic needle which always points north.

CONDUCTOR: a material that allows electricity to flow through it.

CURRENT: a flow of electricity along a wire or some other conductor.

DYNAMO: a machine for generating electricity.

ELECTRICITY: a form of energy used for lighting, heating, and working machinery.

ELECTROMAGNET: an iron bar, surrounded by a coil of wire, which acts like a magnet when electricity flows through the wire.

FILAMENT: a very thin wire in a light bulb that glows white hot when an electric current passes through it.

FUSE: a device that contains a thin wire which heats up and melts if too strong an electric current is passed through it. This breaks the electric circuit and stops the flow of electricity.

GENERATOR: a machine that generates electricity.

INSULATOR: a material that does not allow electricity to flow through it.

LODESTONE: a piece of brown rock that contains iron. It acts as a magnet. It is often called magnetite.

MAGNET: a piece of iron or steel that can attract smaller pieces of iron or steel.

MAGNETIC FIELD: the space around a magnet where the magnetic effects can be felt.

MAGNETIC FORCE: the force that attracts two or more objects to each other.

MAGNETISM: the property of iron, steel, and some other metals to attract or repel a piece of iron.

MAGNETIZED: an object that becomes magnetic after being near a magnet is said to have been magnetized.

POLE: one of the two ends of a magnet. The North Pole and South Pole are the two points on the Earth's surface farthest away from the equator.

POWER PLANT: a large building where electricity is generated.

REPEL: the opposite of attract. To push or turn away from something.

RESISTANCE: something that acts against the flow of electricity, slowing it down.

STATIC ELECTRICITY: a form of electricity that builds up on the surface of materials that do not conduct electricity. It is caused by rubbing.

SWITCH: a device used to start or stop the flow of electricity.

TERMINAL: a point on a battery, bulb holder, or some other part of an electrical circuit where a connection can be made.

TURBINE: a large fan which turns when steam or water is passed over it. This in turn drives a generator.

UTILITY POLE: a tall metal tower used for carrying overhead electric cables.